A COMPARISON BETWEEN

2024 Honda CR-V Hybrid

vs.

Toyota Grand Highlander Hybrid

Your In-depth Car Reviews and Guide Helping you make a choice

Jason J. Berman

Dedication

To the unwavering support and boundless love of my parents who have been the steadfast pillars upon which my dreams have soared. Your encouragement and belief in my journey have been the driving force behind every word penned in this book.

To my well-wishers, whose optimism and cheering voices have echoed in the background of my writing days, providing the motivation needed to overcome challenges and embrace triumphs. Your belief in my creative endeavors has been a source of inspiration that I carry with pride.

And, above all, to the Almighty, the Divine orchestrator of destinies, who granted me the gift of creativity and the opportunity to weave stories. In moments of solitude and uncertainty, your guidance has been my compass, steering me through the labyrinth of imagination.

May the words within these pages be a humble offering of gratitude to those whose love and blessings have illuminated my path. This book is dedicated to the enduring spirits of family, friendship, and the Divine forces that shape our narratives.

Table of Contents

Introduction: Navigating the Hybrid SUV Maze: CR-V vs. Grand Highlander

The automotive landscape is shifting like desert sands, and the future of fuel-efficient driving lies in the burgeoning realm of hybrid SUVs. These sleek machines bridge the gap between gas-guzzling behemoths and eco-conscious econoboxes,offering the practicality of an SUV with the conscience of a hybrid. But with a dizzying array of options, choosing the right one can feel like deciphering a Mayan calendar. Enter this book, your trusty compass through the tangled undergrowth of the 2024 Honda CR-V

Hybrid and Toyota Grand Highlander Hybrid.

Purpose of the Book: Unveiling the Champion

This book is your indispensable guide to navigating the head-to-head battle between two of the most coveted hybrid SUVs on the market. We delve deep into the strengths and weaknesses of each, meticulously dissecting their specs,performance, features, and ownership experience. Our aim is to empower you, the discerning driver, to make an informed decision that aligns perfectly with your needs and desires. Gone are the days of agonizing over brochure numbers and online reviews; this book is your personalized decoder ring, unlocking the secrets of each hybrid SUV and revealing the ultimate champion for your journey.

Overview of the Hybrid SUV Market: A Booming Landscape

The hybrid SUV market is a vibrant ecosystem, teeming with contenders vying for your attention. Fuel efficiency concerns, environmental awareness, and technological advancements have propelled these versatile vehicles into the spotlight. Yet, the sheer variety can be overwhelming. From compact crossovers to spacious three-row behemoths, each hybrid SUV boasts unique selling points and caters to specific lifestyles. This book focuses on two titans of the midsize segment, the

Honda CR-V Hybrid and the Toyota Grand Highlander Hybrid, two vehicles that cater to a broad spectrum of drivers.

The Enduring Allure of the CR-V: For over two decades, the CR-V has been synonymous with practicality and reliability. Its hybrid iteration brings fuel-sipping efficiency to the equation, while its compact dimensions and agile handling make it a city dweller's dream.

The Grand Ambitions of the Grand Highlander: A newcomer to the scene, the Grand Highlander arrives with a bolder presence and spaciousness. Its hybrid powertrain promises both fuel efficiency and the muscle to handle family adventures and weekend getaways.

Target Audience: Who We're Geared Up For

This book speaks to a diverse tribe of drivers united by a desire for a smarter SUV. Whether you're a:

- **Eco-conscious commuter:** Prioritizing fuel efficiency and minimizing environmental impact.
- **Adventure-seeking family:** Craving a spacious and comfortable ride for weekend escapes.
- **Tech-savvy enthusiast:** Demanding cutting-edge features and a seamless driving experience.

- **Budget-minded buyer:** Seeking the best value proposition without compromising quality.
- **Safety-first driver:** Placing utmost importance on advanced safety technologies and driver-assistance features.

This book caters to your individual needs and preferences, ensuring you find the perfect hybrid SUV match. No matter your driving style or priorities, we'll equip you with the knowledge and insights to navigate the CR-V vs. Grand Highlander conundrum with confidence.

Chapter 1: Head-to-Head Comparison: CR-V vs. Grand Highlander

Deciding between the 2024 Honda CR-V Hybrid and Toyota Grand Highlander Hybrid can feel like choosing between seasoned veterans and ambitious newcomers. Both boast impressive attributes, but understanding their individual strengths and weaknesses is crucial for making the right call. Let's dive into a comprehensive head-to-head comparison,focusing on key specifications that define their driving experience, practicality, and technological prowess.

Engine and Performance:

- **CR-V Hybrid:** Equipped with a 2.0L Atkinson-cycle engine paired with an electric motor, generating a combined 215 horsepower. Delivers a spirited but economical drive with responsive acceleration, particularly in urban settings.

- **Grand Highlander Hybrid:** Packs a punch with its 2.5L four-cylinder engine and two electric motors, churning out a robust 295 horsepower. Offers effortless power for highway cruising and tackling inclines, ideal for larger families and adventurous outings.

Fuel Efficiency and Driving Range:

- **CR-V Hybrid:** Takes the crown with an EPA-estimated 40 mpg in the city and 38 mpg on the highway, translating to roughly 510 miles of range before needing a refill. Ideal for budget-conscious drivers and those who prioritize minimizing environmental impact.
- **Grand Highlander Hybrid:** While delivering impressive power, it sacrifices slightly in fuel efficiency. Expect 27 mpg in the city and 28 mpg on the highway, roughly translating to 370 miles on a full tank. Still commendable for its size and power output.

Dimensions and Cargo Space:

- **CR-V Hybrid:** Remains a compact and nimble option, measuring 180.5 inches long and offering 39.2 cubic feet of cargo space behind the rear seats. Suitable for smaller families and those who appreciate maneuverability in tight spaces.
- **Grand Highlander Hybrid:** Embraces spaciousness, stretching 194.9 inches in length and boasting a cavernous 61.1 cubic feet of cargo space with the rear seats up. Perfect for larger families, frequent road trips, and those who demand maximum versatility.

Passenger Capacity and Comfort:

- **CR-V Hybrid:** Comfortably seats five adults with ample headroom and legroom in the front. The rear seats offer good legroom for children or smaller adults, though taller individuals might find it snug.

- **Grand Highlander Hybrid:** Caters to larger families with three seating rows, accommodating seven passengers.The first two rows provide ample space for adults, while the third row is better suited for children or occasional use.

Safety Features and Technology:

- **CR-V Hybrid:** Comes standard with the Honda Sensing suite, including automatic emergency braking, lane departure warning, and adaptive cruise control. Available features include a surround-view camera and head-up display.

- **Grand Highlander Hybrid:** Equipped with Toyota Safety Sense 3.0, offering similar active safety features like pre-collision braking and lane departure warning, along with blind-spot monitoring and rear cross-traffic alert.Higher trims introduce advanced features like a panoramic moonroof and a digital instrument cluster.

Summary:

The CR-V Hybrid excels in fuel efficiency, maneuverability, and city driving, prioritizing eco-conscious commuters and smaller families. The Grand Highlander Hybrid counters with its spaciousness, robust power, and advanced technology,catering to larger families and adventure seekers. Choosing the right one depends on your needs, priorities, and driving habits. Weigh the advantages and disadvantages carefully, taking test drives to gauge comfort and performance firsthand.Remember, the perfect hybrid SUV is one that seamlessly integrates with your lifestyle, delivering optimal functionality and driving satisfaction.

Chapter 2: Deep Dive: Unveiling the Nuances of CR-V vs. Grand Highlander

Beneath the surface of impressive specifications lies a world of nuanced details that truly differentiate the 2024 Honda CR-V Hybrid and Toyota Grand Highlander Hybrid. Let's dive deeper into each aspect, unveiling the subtle strengths and quirks that might sway your decision towards one or the other.

Performance and Handling:

- **CR-V Hybrid:** Excels in urban environments with its peppy acceleration and nimble handling. The electric motor provides instant torque, making city commutes a breeze. However, the continuously variable transmission (CVT) can feel rubbery at higher speeds, and steering lacks some feedback. Overall, a comfortable and efficient ride, but not exhilarating.

- **Grand Highlander Hybrid:** Delivers surprisingly sporty handling for its size, thanks to its lower center of gravity and well-tuned suspension. The powerful engine offers confident highway passing and effortless hill climbing.However, the larger dimensions make it less manoeuvrable in tight spaces. Opt for higher trims for improved steering feel and adaptive dampers for superior ride quality.

Fuel Economy and Emissions:

- **CR-V Hybrid:** Takes the undisputed crown with its outstanding fuel efficiency. Expect 40 mpg in the city and 38 mpg on the highway, translating to an impressive 510-mile range. Its EPA-estimated annual fuel cost sits below $1,000, making it a budget-friendly option. Low CO_2 emissions make it a responsible choice for the environment.

- **Grand Highlander Hybrid:** While less fuel-efficient than the CR-V, it performs admirably for its size and power.Expect 27 mpg in the city and 28 mpg on the highway, offering roughly 370 miles of range. Higher trims with all-wheel-drive see a slight decrease in fuel economy. Nonetheless, the hybrid powertrain significantly reduces emissions compared to its purely gasoline-powered counterparts.

Interior Comfort and Features:

- **CR-V Hybrid:** Offers a comfortable and well-designed cabin with ample headroom and legroom for the front passengers. The rear seats are spacious for children but might feel snug for taller adults. Soft-touch materials and subtle design elements create a refined ambience. Higher trims boast heated seats, panoramic sunroof, and a premium sound system.

- **Grand Highlander Hybrid:** Impresses with its expansive interior, comfortably accommodating seven passengers.The spacious second row and surprisingly usable third row make it ideal for families. Higher trims offer luxury features like leather seats, ventilated front seats, and a heated steering wheel. However, the overall design feels less refined than the CR-V, with some hard plastics and less intuitive ergonomics.

Technology and Infotainment:

- **CR-V Hybrid:** Features Honda's user-friendly infotainment system with a responsive touchscreen, Apple CarPlay,Android Auto, and optional navigation. Additional features like a head-up display and wireless charging are available depending on the trim level. However, the system can feel slightly dated compared to some competitors.

- **Grand Highlander Hybrid:** Boasts Toyota's latest infotainment system with a larger touchscreen, faster processing, and improved graphics. Apple CarPlay and Android Auto are standard, along with optional features like a panoramic moonroof and a digital instrument cluster. The system feels more modern and intuitive than the CR-V's, but voice recognition can be occasionally frustrating.

Safety Ratings and Features:

- **CR-V Hybrid:** Earns a five-star overall safety rating from the National Highway Traffic Safety Administration (NHTSA) and a "Top Safety Pick+" designation from the Insurance Institute for Highway Safety (IIHS). Standard safety features include automatic emergency braking, lane departure warning, and adaptive cruise control. Higher trims offer features like blind-spot monitoring and surround-view camera.

- **Grand Highlander Hybrid:** Receives a five-star overall safety rating from NHTSA and is currently being evaluated by IIHS. Toyota Safety Sense 3.0 suite comes standard, offering similar active safety features as the CR-V. Higher trims add advanced features like blind-spot monitoring with cross-traffic alert, rear-cross-traffic alert, and a panoramic view monitor.

Summary:

Choosing between the CR-V Hybrid and Grand Highlander Hybrid is a nuanced decision. The CR-V excels in fuel efficiency, interior design, and infotainment, making it ideal for eco-conscious urban dwellers and smaller families. The Grand Highlander shines with its spaciousness, power, and safety features, perfect for larger families and adventurous outings. Ultimately, the choice boils down to your individual needs and priorities. Consider your driving habits, passenger needs, technological preferences, and budget to determine which hybrid SUV resonates with your lifestyle.

Chapter 3: Strengths and Weaknesses: Unveiling the True Colors of CR-V and Grand Highlander

The 2024 Honda CR-V Hybrid and Toyota Grand Highlander Hybrid stand tall in the hybrid SUV arena, but beneath their shiny exteriors lie unique strengths and weaknesses waiting to be unearthed. Identifying these hidden gems and potential pitfalls is crucial for choosing the champion that perfectly complements your driving journey.

CR-V Hybrid: The Eco-Savvy Ace with a Compact Appeal

Strengths:

- **Fuel Efficiency King:** With an EPA-estimated 40 mpg in the city and 38 mpg on the highway, the CR-V reigns supreme in fuel economy. This translates to a budget-friendly and environmentally conscious ride, ideal for city dwellers and frequent commuters.

- **Nimble Urban Warrior:** Its compact dimensions and responsive handling make navigating city streets a breeze.Smooth steering and a tight turning radius ensure effortless parking and maneuvering through tight spaces.

- **Cabin Comfort and Refinement:** The CR-V boasts a

thoughtfully designed and well-built interior. Soft-touch materials, an intuitive infotainment system, and optional features like heated seats and panoramic sunroof create a comfortable and refined ambience.

- **Safety Standard Bearer:** Earning a five-star NHTSA rating and "Top Safety Pick+" designation from IIHS, the CR-V prioritizes passenger safety with a comprehensive suite of active safety features like automatic emergency braking and lane departure warning.

Weaknesses:

- **Uninspiring Performance:** While adequate for city driving, the CR-V's CVT and 215 horsepower don't offer thrilling acceleration or confident highway passing. Expect a leisurely driving experience focused on efficiency over excitement.

- **Space Limitations:** The compact size translates to slightly cramped rear seats for taller passengers. Large families or those who prioritize cargo space might find the Grand Highlander a better fit.

- **Dated Technology:** While functional, the CR-V's infotainment system feels slightly dated compared to its competitor's offerings. Features like a digital instrument cluster or advanced voice recognition are absent.

- **Higher Ownership Costs:** While fuel-efficient, the CR-V's starting price might be higher than some competitors.Additionally, optional features like all-wheel-drive and advanced safety technologies can further inflate the cost.

Grand Highlander Hybrid: Spacious Powerhouse with Grand Aspirations

Strengths:

- **Commanding Presence and Power:** The Grand Highlander's larger size and 295 horsepower engine offer a confident and powerful driving experience. It tackles highway cruising and handles inclines with ease, making it ideal for long road trips and adventures.

- **Family-Friendly Haven:** The spacious interior comfortably accommodates seven passengers, with ample headroom and legroom even in the third row. This makes it a haven for large families and frequent carpoolers.

- **Advanced Technology Playground:** The Grand Highlander boasts Toyota's latest infotainment system featuring a larger touchscreen, faster processing, and optional features like a panoramic moonroof and digital instrument cluster. Technological enthusiasts will find it a more modern and engaging experience.

- **Uncompromising Safety:** Receiving a five-star NHTSA rating and awaiting IIHS evaluation, the Grand Highlander offers a comprehensive suite of safety features to ensure passenger protection on every journey.

Weaknesses:

- **Fuel Efficiency Compromise:** While impressive for its size and power, the Grand Highlander's 27 city mpg and 28 highway mpg fall short of the CR-V's stellar fuel efficiency. Higher fuel costs and environmental impact are trade-offs for spaciousness and power.

- **Maneuverability Challenges:** The larger size can feel clunky in tight spaces. City driving and parking might require more planning and adjustment compared to the nimble CR-V.

- **Interior Design Quirks:** While spacious, the Grand Highlander's interior design elements and material choices might feel less refined than the CR-V's, with some hard plastics and less intuitive ergonomics.

- **Price Premium:** The Grand Highlander's starting price is higher than the CR-V, and higher trims with advanced features can significantly drive up the cost. Consider your budget carefully before choosing the perfect trim level.

Summary:

The CR-V Hybrid and Grand Highlander Hybrid paint distinct portraits of efficiency, performance, and practicality.Choose the CR-V for its unmatched fuel economy, nimble handling, and refined interior, ideal for eco-conscious commuters and smaller families. Opt for the Grand Highlander for its spaciousness, powerful engine, and advanced technology, perfect for large families and adventure seekers.

Ultimately, the champion lies in the one that best complements your lifestyle and driving needs. Weigh your priorities, understand the strengths and weaknesses, and let your journey begin with the perfect hybrid SUV by your side.

Chapter 4: Counting Pennies and Miles: The CR-V and Grand Highlander's Financial Frontier

In the grand dance of choosing your ideal hybrid SUV, the music inevitably turns to the rhythm of cost. While both the 2024 Honda CR-V Hybrid and Toyota Grand Highlander Hybrid boast stellar features, understanding their pricing strategies and ownership expenses is crucial for a financially harmonious relationship.

Pricing Range: Mapping the Trim Level Landscape

CR-V Hybrid:

- **LX:** Starting at $28,245, this base trim offers the fuel-efficient hybrid powertrain, standard safety features, and a basic infotainment system.
- **EX:** At $31,525, you get heated seats, a moonroof, and an upgraded audio system.
- **EX-L:** Priced at $34,305, this trim adds leather seats, blind-spot monitoring, and rear cross-traffic alert.
- **Touring:** For $37,705, you gain a navigation system, heated steering wheel, and adaptive cruise control.
- **Touring Hybrid:** The top trim at $40,405 comes with premium audio, wireless charging, and a surround-view camera.

Grand Highlander Hybrid:

- **Base:** Starting at $36,120, this base trim offers the hybrid powertrain, standard safety features, and a basic infotainment system.
- **LE:** At $38,920, you get heated seats, a sunroof, and an upgraded audio system.
- **XLE:** Priced at $42,720, this trim adds leather seats, blind-spot monitoring, and rear cross-traffic alert.
- **Limited:** For $47,720, you gain a navigation system, heated steering wheel, and adaptive cruise control.
- **Platinum:** The top trim at $52,720 features a premium audio

system, wireless charging, and a panoramic moonroof.

Option Costs: Adding Flavor to Your Ride

Both SUVs offer a range of optional features across different trims. Be prepared to add thousands to the base price if you crave luxuries like advanced driver-assistance systems, panoramic roofs, and premium audio upgrades. Remember, each option chips away at your budget, so prioritize wisely based on your needs and desires.

Fueling the Flames: A Tale of Two Tanks

The CR-V's stellar fuel economy shines in this battle. Expect to pay roughly

$1,000 annually for gas compared to the Grand Highlander's estimated $1,300, based on average US gas prices and driving habits. While the Grand Highlander offers more power and space, the CR-V delivers significant savings at the pump, especially for frequent commuters and city dwellers.

Maintenance Matters: Keeping Your Hybrid Humming

Both SUVs boast reputations for reliability and minimal maintenance costs. Expect standard hybrid battery and electric motor warranties in addition to the regular powertrain warranty. Both have fairly similar maintenance costs, roughly $500-$800 annually for routine servicing. However, remember that premium features like adaptive dampers or advanced driver-assistance systems might have higher repair costs down the line.

Insurance Interlude: Protecting Your Investment

Insurance premiums for both hybrid SUVs will vary depending on your personal driving history, location, and chosen trim level. The CR-V's smaller size and fuel efficiency might translate to slightly lower insurance costs compared to the Grand Highlander's larger stature and powerful engine. Remember to compare quotes from different insurance companies before signing on the dotted line.

Summary: Choosing Your Financial Champion

The cost dance between the CR-V Hybrid and Grand Highlander Hybrid comes down to your priorities and driving habits.For budget-conscious and eco-minded drivers, the CR-V's lower starting price, superior fuel economy, and slightly lower insurance costs make it a clear victor. The Grand Highlander, however, tempts with its spaciousness, power, and advanced technology, though these come at the cost of a higher base price, slightly lower fuel efficiency, and potentially higher insurance.

Ultimately, the champion is the one that best balances your financial constraints with your desire for features and functionality. Choose wisely, count your pennies carefully, and prepare to embark on a financially rewarding journey with your perfect hybrid SUV companion.

Chapter 5: Critical Analysis into the 2024 Honda CR-V Hybrid: A Modern Evolution of an Evergreen Favorite

The Honda CR-V needs no introduction. For over two decades, it's reigned supreme as a practical and reliable SUV,beloved by families and commuters alike. The 2024 Hybrid iteration adds a fuel-efficient twist to this evergreen favorite,promising the same practicality with a greener conscience. Let's delve into the heart of this hybrid evolution, examining its history, driving experience, technology, and how it stacks up against rivals.

From Humble Beginnings to Hybrid Hero:

The CR-V's story began in 1995, capturing hearts with its compact size and spacious interior. Each generation refined the formula, adding features, improving performance, and solidifying its reputation for dependability. Today, the fifth-generation CR-V Hybrid carries the torch, embracing electrification while staying true to its practical roots.

Behind the Wheel: A Refined Balance of Efficiency and Fun

The CR-V Hybrid's 2.0L Atkinson-cycle engine and electric motor combo deliver 215 horsepower, promising a peppy yet economical ride. The continuously variable transmission (CVT) ensures smooth acceleration, particularly in urban settings. While not a speed demon, the CR-V handles daily commutes with ease, offering responsive steering and decent agility for navigating tight city streets.

Stepping Inside: A Cabin of Comfort and Convenience

The CR-V Hybrid's interior is a well-designed haven. Soft-touch materials, ergonomic controls, and ample headroom and legroom create a comfortable environment for all passengers. The infotainment system features a responsive touchscreen with Apple CarPlay and Android Auto, facilitating seamless smartphone integration. Higher trims offer additional luxuries like heated seats, a panoramic sunroof, and a premium sound system, elevating the experience further.

Technology that Impresses, Not Overwhelms:

The CR-V Hybrid's technology package focuses on user-friendliness. The infotainment system is intuitive and straightforward, providing all essential functions without feeling cluttered. Honda Sensing, a suite of active safety features like automatic emergency braking and lane departure warning, comes standard, ensuring peace of mind on every journey.Advanced features like a head-up display and surround-view camera are available on higher trims.

Facing the Competition: Where the CR-V Hybrid Shines

The CR-V Hybrid faces stiff competition from established rivals like the Toyota RAV4 Hybrid and newcomers like the Ford Escape Hybrid. Compared to the RAV4, the CR-V boasts slightly better fuel economy and a more refined interior.The Escape Hybrid offers a sportier driving experience, but falls short in interior space and cargo capacity. Ultimately, the CR-V Hybrid stands out with its exceptional fuel efficiency, comfortable and well-equipped cabin, and user-friendly technology, making it a compelling choice for eco-conscious drivers seeking a practical and reliable SUV.

The 2024 Honda CR-V Hybrid skillfully blends its heritage of practicality with the modern appeal of hybrid technology.

It delivers a smooth and efficient driving experience, a comfortable and well-designed interior, and intuitive technology, all while remaining budget-friendly. While other offerings might excel in specific areas, the CR-V Hybrid excels at striking a perfect balance, making it a worthy champion for families, commuters, and anyone seeking a reliable and fuel-efficient SUV that doesn't compromise on comfort or features.

Chapter 6: Grand Ambitions: Unveiling the 2024 Toyota Grand Highlander Hybrid

Toyota, the titan of hybrid SUVs, throws down the gauntlet with a newcomer in 2024: the Grand Highlander Hybrid. This behemoth aims to conquer the spaciousness segment, carving a niche between the trusty Highlander and the Sequoia behemoth. Let's embark on a deep dive, exploring the Grand Highlander's driving dynamics, comfort, entertainment, and its unique position within Toyota's lineup.

A Giant Leap in Grandeur:

The Grand Highlander Hybrid arrives as the missing link in Toyota's SUV chain. Bigger than the Highlander but smaller than the Sequoia, it caters to families yearning for an extra dose of space and power without venturing into full-size territory. Its hybrid powertrain promises not only ample room but also eco-conscious efficiency, a hallmark of Toyota's engineering prowess.

Taming the Road with Power and Grace:

The Grand Highlander Hybrid packs a punch with its 2.5L four-cylinder engine and two electric motors, churning out a robust 295 horsepower. This translates to effortless highway cruising and confident hill climbing, perfect for adventurous families and frequent road trippers. While its size might suggest sluggishness, the Grand Highlander surprises with surprisingly nimble handling thanks to its well-tuned suspension and lower center of gravity. However, navigating tight city streets might require more planning compared to its smaller counterparts.

A Haven of Comfort for Seven:

Inside, the Grand Highlander Hybrid is a spacious oasis. Three rows of seats comfortably accommodate seven passengers,with ample headroom and legroom even for adults in the third row. Higher trims elevate the experience with features like heated and ventilated seats, a panoramic moonroof, and a digital instrument cluster. While some hard plastics are present,the overall design feels more upscale than the CR-V

Hybrid, with a focus on comfort and functionality.

Entertainment Evolved: Keeping Everyone Engaged:

Toyota's latest infotainment system shines in the Grand Highlander Hybrid. A responsive touchscreen displays crisp graphics and offers features like Apple CarPlay, Android Auto, and optional navigation. Higher trims boast features like a premium sound system and a rear-seat entertainment system, ensuring everyone stays entertained on long journeys. While the voice recognition system can be occasionally frustrating, the overall experience is intuitive and user-friendly.

Within the Toyota Fold: Grand Highlander's Market Niche:

The Grand Highlander Hybrid fills a crucial gap in Toyota's hybrid SUV lineup. Compared to the Highlander Hybrid, it offers significantly more space and a more powerful engine, while remaining fuel-efficient and family-friendly. Compared to the Sequoia, it delivers better fuel economy, a more manageable size, and a hybrid option, making it more appealing for environmentally conscious families. The Grand Highlander Hybrid, therefore, carves its own niche, catering to large families who prioritize spaciousness, power, and a touch of eco-consciousness, without venturing into full-size SUV territory.

The 2024 Toyota Grand Highlander Hybrid is a bold statement, offering space, power, and family-friendly features wrapped in a hybrid powertrain. Its driving dynamics are surprisingly nimble for its size, while the interior is a haven of comfort and entertainment. Within Toyota's lineup, it fills a crucial gap, appealing to families who yearn for more than the Highlander Hybrid can offer but find the Sequoia too daunting. While fuel efficiency might not be its strongest suit, the Grand Highlander Hybrid delivers a compelling combination of space, power, and eco-consciousness, making it a strong contender in the growing market of large hybrid SUVs.

Chapter 7: From City Streets to Mountain Passes: Matching the Hybrid SUV to Your Lifestyle

Beyond the glitzy specs and dazzling features lies the true test of any car: your everyday life. Choosing the right hybrid SUV isn't just about horsepower and cargo space; it's about finding the perfect match for your lifestyle, needs, and driving habits. Let's dissect the ownership experience of the 2024 Honda CR-V Hybrid and Toyota Grand Highlander Hybrid,identifying their ideal use cases and target buyers.

CR-V Hybrid: The Urban Eco-Warrior's Ally

- **Target Buyers:** Eco-conscious city dwellers, commuting families, solo adventurers
- **Ideal Use Cases:** Daily commutes, city errands, weekend getaways, light off-roading
- **Strengths:** Superior fuel efficiency, compact size, nimble handling, user-friendly technology, affordability
- **Weaknesses:** Limited cargo space, cramped rear seats for taller passengers, less powerful than the Grand Highlander
- **Verdict:** The CR-V Hybrid champions urban living and efficient exploration. Its fuel-sipping nature makes it ideal for daily commutes, while its

compact size and agile handling make navigating city streets a breeze. Smaller families and solo adventurers will appreciate its practicality and comfortable cabin, while its light off-road capabilities make weekend escapes feasible. However, larger families seeking ample space or frequent road trippers might find its limitations frustrating.

Grand Highlander Hybrid: The Spacious Adventurer's Haven

- **Target Buyers:** Large families, frequent road trippers, outdoor enthusiasts

- **Ideal Use Cases:** Family road trips, long journeys, camping trips, occasional city driving
- **Strengths:** Spacious and comfortable interior, powerful engine, advanced technology, diverse trim options
- **Weaknesses:** Lower fuel efficiency than the CR-V, larger size might be cumbersome in tight spaces, higher starting price
- **Verdict:** The Grand Highlander Hybrid embraces the open road with open arms. Its cavernous interior comfortably accommodates seven passengers, making it ideal for large families and frequent road trips. The powerful engine tackles highways

and inclines with ease, while advanced technology keeps everyone entertained and connected.However, its larger size might require more planning in urban environments, and its lower fuel efficiency might deter city dwellers who prioritize eco-conscious commuting.

Beyond the Numbers: Considering Practicalities

Owning a hybrid SUV goes beyond fuel efficiency and horsepower. Consider these practical aspects when making your choice:

- **Budget:** Both offer a range of trims, but the Grand Highlander's starting price is higher. Choose wisely based on your needs and budget constraints.

- **Fuel Costs:** While the CR-V shines in fuel efficiency, the Grand Highlander's lower mpg might translate to higher costs, especially for frequent drivers.
- **Maintenance:** Both boast reliability, but consider potential repair costs of advanced features like adaptive dampers or complex infotainment systems.
- **Insurance:** The Grand Highlander's size and power might result in slightly higher insurance premiums compared to the CR-V.

Matching Your Lifestyle to the Perfect Hybrid SUV

Imagine your typical driving experience. Do you navigate city streets daily, or do you yearn for weekend escapes into the wild? Are you a solo adventurer or a family of adventurers? Your answers will guide you towards the perfect hybrid SUV:

- **City Dweller:** Choose the CR-V Hybrid for its fuel efficiency, nimble handling, and compact size.

- **Eco-Conscious Commuter:** Both are great options, but the CR-V edges out with its superior mpg.

- **Family of Four:** Both work well, but the CR-V might feel cramped for extended trips. Consider the

Grand Highlander for more space and comfort.

- **Large Family/Frequent Road Trippers:** The Grand Highlander's spaciousness, power, and advanced technology make it the ideal companion.
- **Outdoor Enthusiast:** Both can handle light off-roading, but the CR-V's size might give it an edge on tight trails.

Ultimately, the perfect hybrid SUV is the one that seamlessly integrates with your lifestyle. Test drive both vehicles,consider your practical needs, and trust your gut instinct. With this guide and a thoughtful approach, you'll be cruising towards a journey filled with adventure, efficiency, and the perfect hybrid SUV by your side.

Chapter 8: Beyond the Showroom: Living with the CR-V and Grand Highlander Hybrid

Choosing a hybrid SUV isn't just about a shiny brochure and test drive; it's about welcoming a new companion into your daily life. The 2024 Honda CR-V Hybrid and Toyota Grand Highlander Hybrid, while sharing the hybrid mantle, cater to distinct lifestyles and expectations. Let's delve into the nitty-gritty of ownership, identifying their ideal use cases and the types of drivers who'll thrive behind the wheel.

CR-V Hybrid: The Urban Eco-Champion's Chariot

- **Target Buyers:** Eco-conscious city dwellers, solo adventurers, families with younger children
- **Ideal Use Cases:** Daily commutes, errands within the city, weekend camping trips, light hikes
- **Strengths:** Superior fuel efficiency, compact size, nimble handling, user-friendly technology, affordability
- **Weaknesses:** Limited cargo space, cramped rear seats for taller passengers, less powerful than the Grand Highlander
- **Daily Life:** Imagine weaving through city streets, effortlessly finding parking, and relishing the smug satisfaction of your fuel gauge barely budging. The CR-V

Hybrid thrives in urban choreography, its small size a blessing in tight spaces and its fuel efficiency a balm for the eco-conscious soul. Solo adventurers and smaller families will enjoy its comfortable cabin and practical features, while its light off-road capabilities make weekend escapes to nearby trails doable. However, larger families or frequent road trippers might find its space limitations and lack of power frustrating.

Grand Highlander Hybrid: The Spacious Adventurer's Fortress

- **Target Buyers:** Large families, frequent road trippers, outdoor enthusiasts
- **Ideal Use Cases:** Family road trips, long journeys, camping ventures, exploring diverse terrains
- **Strengths:** Spacious and comfortable interior, powerful engine, advanced technology, diverse trim options
- **Weaknesses:** Lower fuel efficiency than the CR-V, larger size might be cumbersome in tight spaces, higher starting price
- **Daily Life:** Picture cruising down highways with the expansive vista spread before you, seven passengers comfortably

ensconced, technology keeping everyone entertained, and the power to tackle any incline. The Grand Highlander Hybrid embraces the open road, its cavernous interior a haven for large families and long journeys. The powerful engine effortlessly conquers highways and mountain passes, while advanced driver-assistance features ensure confident navigation on diverse terrains. However, city dwellers might find its size inconvenient, and its lower fuel efficiency might not appeal to those prioritizing eco-conscious commuting.

Beyond the Specs: Practicalities Play a Role

Owning a hybrid SUV goes beyond the technical dance of horsepower and mpg. Consider these practical aspects when making your choice:

- **Budget:** Both offer a range of trims, but the Grand Highlander's starting price is higher. Choose wisely based on your needs and financial constraints.
- **Fuel Costs:** While the CR-V shines in fuel efficiency, the Grand Highlander's lower mpg might translate to higher costs, especially for frequent drivers.
- **Maintenance:** Both boast reliability, but consider potential repair costs of advanced features like adaptive dampers or complex infotainment systems.

- **Insurance:** The Grand Highlander's size and power might result in slightly higher insurance premiums compared to the CR-V.

Lifestyle Matches: Finding Your Perfect Hybrid SUV

Imagine your typical driving experience. Do you navigate city streets daily, or do you yearn for weekend escapes into the wild? Are you a solo adventurer or a family of adventurers? Your answers will guide you towards the perfect hybrid SUV:

- **City Dweller:** Choose the CR-V Hybrid for its fuel efficiency, nimble handling, and compact size.

- **Eco-Conscious Commuter:** Both are great options, but the CR-V edges out with its superior mpg.

- **Family of Four:** Both work well, but the CR-V might feel cramped for extended trips. Consider the Grand Highlander for more space and comfort.

- **Large Family/Frequent Road Trippers:** The Grand Highlander's spaciousness, power, and advanced technology make it the ideal companion.

- **Outdoor Enthusiast:** Both can handle light off-roading, but the CR-V's size might give it an edge on tight trails.

Life with your Hybrid SUV: A Tailored Experience

Remember, ownership is a personal journey. Don't be afraid to personalize your book by including:

- **Real-life ownership experiences:** Let others paint a picture of daily life with each hybrid SUV.
- **Detailed cost breakdowns:** Help readers understand the financial implications of ownership.
- **Decision-making tools:** Quizzes, comparison charts, and pro-con lists can empower readers to make informed choices.

Chapter 9: Beyond the Brochure: Test Driving Your Way to the Perfect Hybrid SUV

Stepping into a showroom, brochures gleaming, you're faced with a pivotal decision: choosing the ideal hybrid SUV. While specifications and reviews offer valuable insights, it's the test drive that truly seals the deal. But approaching a test drive blindly can be a recipe for disappointment. Fear not, intrepid explorer, for this guide equips you with the tools to ace your hybrid SUV test drive and find the perfect match for your unique needs and preferences.

Know Thyself: Prioritizing Features for a Flawless Fit

Before hopping behind the wheel, take a moment of introspection. Ask yourself:

- **City Dweller or Open Road Wanderer?** Do you navigate city streets daily or crave weekend escapes into the vast unknown? Fuel efficiency and maneuverability might be paramount for one, while power and spaciousness for long journeys reign supreme for the other.

- **Tech Whiz or Analogue Hero?** Do cutting-edge infotainment systems and advanced driver-assistance features excite you, or do you prefer a simpler, less distracting experience? Prioritize features that enhance your driving comfort and avoid ones that feel unnecessary or overwhelming.

- **Solo Adventurer or Family Flock?** Consider how often you'll transport passengers and how much cargo space you require. A compact size might be ideal for solo pursuits, while a roomy interior and powerful engine become crucial for family adventures.

- **Budgetary Bard or Financially Free?** Price is a crucial factor. Determine your budget before test driving to avoid falling in love with a vehicle beyond your means.

Taking the Wheel: A Symphony of Senses

As you grip the steering wheel, let your senses become your guide:

- **Sight:** Survey the interior. Does the layout feel intuitive? Are the materials comfortable and appealing? Do the controls fall readily to hand?
- **Touch:** Run your fingers across the surfaces. Are they soft and supple, or rough and plasticky? Does the steering wheel feel comfortable and responsive?
- **Sound:** Listen for engine noise, wind buffeting, and road imperfections. Is the cabin well-insulated, or does every bump become a symphony of rattles?
- **Smell:** Does the interior have a fresh, pleasant scent, or does it

carry an unpleasant chemical odor?

- **Feel:** This is where the magic happens. Ease onto the accelerator. Does the car respond smoothly or hesitantly? Take a corner. Does it handle with agility or feel cumbersome? Hit the brakes. Are they responsive and reassuring, or do they feel spongy and inspire uncertainty?

Beyond the Basics: Diving Deeper for Dealbreakers

Don't be afraid to push the boundaries of your test drive. Here's how:

- **City Streets:** Navigate tight spaces, practice parallel parking, and assess maneuverability in real-world traffic situations.

- **Highway Cruising:** Merge onto the highway, feel the engine's power under acceleration, and evaluate noise levels at higher speeds.

- **Bumpy Roads:** Seek out uneven terrain to assess the suspension's ability to handle imperfections.

- **Advanced Features:** If your chosen trim boasts advanced driver-assistance features like lane departure warning or adaptive cruise control, put them to the test and ensure they operate smoothly and intuitively.

Heeding the Red Flags: Recognizing Dealbreakers

Not every test drive will be a love story. Be on the lookout for dealbreakers:

- **Uncomfortable Ergonomics:** Does the seating position feel awkward? Do you find yourself constantly adjusting?

- **Blind Spots Galore:** Are the visibility limitations a safety concern? Can you navigate confidently with the available mirrors and cameras?

- **Technology Tantrums:** Does the infotainment system lag or glitch? Are the voice commands frustratingly inaccurate?

- **Fuel Frenzy:** Does the fuel gauge plummet at an alarming rate during your test drive? Consider real-world fuel efficiency against your driving habits.

- **Gut Feeling Woes:** Listen to your intuition. Does the car feel like a natural extension of yourself, or are you constantly battling its quirks and idiosyncrasies?

Choosing with Confidence: Your Perfect Hybrid SUV Awaits

By combining self-awareness with a thorough test drive, you'll transform the daunting task of choosing the right hybrid SUV into a fulfilling and empowering experience. Remember, there's no one-size-fits-all solution. Prioritize your needs, engage your senses, and don't hesitate to walk away if something feels off. With this guide as your compass, you'll navigate the test drive journey with confidence, knowing that your perfect hybrid SUV awaits, ready to embark on countless adventures with you at the wheel.

Further Enrichment: Personalize Your Guide

To ensure your readers find the perfect match, consider these additions:

- **Safety Features:** A detailed comparison of the safety features offered by the CR-V Hybrid and Grand Highlander Hybrid, including standard and optional features like airbags, driver-assistance systems, and crash test ratings.

- **Environmental Impact:** An analysis of the fuel efficiency and emissions of each hybrid SUV, exploring not only their tailpipe emissions but also their overall lifecycle impact on the environment.

- **Long-Term Ownership:** A discussion about the potential costs of ownership beyond the initial purchase price,including insurance, maintenance, and resale value.

- **Final Verdict:** A definitive recommendation for each hybrid SUV based on a combined analysis of all the factors discussed in the previous chapters.

- **Personal Anecdotes:** You could include real-life ownership experiences from drivers who have chosen either the CR-V Hybrid or Grand Highlander Hybrid to give your readers a more personal perspective.

Chapter 10: Counting Pennies and Miles: Demystifying the Cost of Owning a Hybrid SUV

Choosing a hybrid SUV isn't just about saving the planet; it's also about saving your wallet. While both the 2024 Honda CR-V Hybrid and Toyota Grand Highlander Hybrid promise eco-friendly efficiency, their ownership costs paint a nuanced picture, influenced by factors like warranty coverage, maintenance needs, and potential repair expenses. Let's delve into the financial frontiers of these two hybrids, equipping you with the knowledge to make a truly informed choice.

Warranty Worries: Peace of Mind in Every Mile

Both SUVs offer comprehensive warranties, providing peace of mind while traversing the miles. Here's a breakdown:

- **Powertrain Warranty:** Both boast a standard 5-year/60,000-mile powertrain warranty, covering critical components like the engine, hybrid system, and transmission.

- **Hybrid Component Warranty:** Toyota offers an additional 8-year/100,000-mile warranty on the battery and related hybrid components, while Honda provides a 3-year/36,000-mile warranty on the same.

- **Basic Warranty:** Both offer a standard 3-year/36,000-mile bumper-to-bumper warranty for coverage against non-powertrain issues.
- **Corrosion Warranty:** Both provide a 5-year/unlimited-mile warranty against rust and perforation.

Service Station Symphony: Keeping Your Hybrid Humming

Regular maintenance is crucial for optimal performance and longevity. Expect similar intervals for both SUVs:

- **Oil Changes:** Hybrids don't require traditional oil changes, but routine engine inspections and fluid replacements occur every 6 months or 5,000 miles.

- **Tire Rotations and Balances:** Every 5,000 miles, ensure optimal tire wear and performance with regular rotations and balances.

- **Hybrid System Checks:** Every 12 months or 15,000 miles, technicians will inspect the hybrid battery and related components for potential issues.

- **Scheduled Maintenance:** Both manufacturers offer scheduled maintenance plans for additional peace of mind,covering the costs of routine services within a set period.

Expected Expenses: Fueling the Financial Forecast

While both boast impressive fuel efficiency, their size and powertrain differences translate to varying costs:

- **Fuel Costs:** The CR-V Hybrid shines with an EPA-estimated 40 mpg combined, translating to roughly $1,000 annual fuel costs (based on average US gas prices and driving habits). The Grand Highlander Hybrid, with its larger size and less impressive 29 mpg combined, might cost around $1,300 annually.

- **Insurance:** The Grand Highlander's size and power might result in slightly higher insurance premiums compared to the CR-V. Consider comparing quotes from different companies for the most accurate estimate.

- **Maintenance Costs:** Regular service costs for both are fairly similar, averaging around $500-$800 annually.However, remember that the Grand Highlander's more complex features like adaptive dampers or advanced driver-assistance systems might have higher repair costs down the line.

The Bottom Line: Counting Your Pennies for the Perfect Fit

The CR-V Hybrid emerges as the budget-friendly option, with lower fuel costs, slightly cheaper insurance, and potentially lower future repair costs. Its smaller size might also make parking and navigating tight spaces easier. However, the Grand Highlander Hybrid's spaciousness, power, and advanced features might be worth the slightly higher investment for larger families and frequent road trippers.

Ultimately, the choice comes down to your individual needs and priorities. Consider:

- **Budget:** How much can you comfortably afford to spend on fuel, insurance, and maintenance?
- **Driving Habits:** Do you primarily navigate city streets or hit the open road frequently?

- **Space Needs:** How much cargo space and passenger room do you require?
- **Feature Preferences:** Which features are most important to you, and are you willing to pay extra for them?

Beyond the Numbers: Personalizing Your Ownership Experience

Personalize your guide by including:

- **Real-life ownership experiences:** Let actual owners share their maintenance costs and unexpected expenses.
- **Fuel cost calculators:** Empower readers to estimate their annual

fuel costs based on their specific driving habits.

- **Maintenance schedule comparisons:** Provide detailed breakdowns of each SUV's recommended service intervals and associated costs.

- **Insurance quote comparisons:** Guide readers towards resources for comparing insurance quotes from different providers.

Conclusion: The Verdict: Choosing Your Champion from the Hybrid Arena

The 2024 Honda CR-V Hybrid and Toyota Grand Highlander Hybrid stand tall as worthy contenders in the hybrid SUV arena. Both possess undeniable strengths, catering to distinct lifestyles and prioritizing different aspects of the driving experience. But under the microscope of a final verdict, only one can ascend the podium as the champion for your individual needs.

The Eco-Conscious City Dweller's Choice: Honda CR-V Hybrid

If city streets are your daily domain, fuel efficiency reigns supreme, and maneuverability is a must, the CR-V

Hybrid emerges as the undisputed ruler. Its compact size makes parking a breeze, its nimble handling conquers tight traffic knots, and its 40 mpg combined fuel economy puts a smile on both your wallet and the environment. The user-friendly technology keeps you connected, while the comfortable cabin provides a haven for daily commutes and weekend escapes. And at a more budget-friendly starting price, the CR-V Hybrid is a tempting proposition for eco-conscious city dwellers and solo adventurers.

The Spacious Adventurer's Dream: Toyota Grand Highlander Hybrid

For those who crave unhindered exploration, embrace family road trips, and demand power under the hood, the Grand Highlander Hybrid roars to life as the ideal companion. Its cavernous interior comfortably accommodates seven passengers, its powerful engine tackles mountain passes with ease, and its advanced technology keeps everyone entertained even on the longest journeys. While its lower fuel efficiency might sting the wallet of eco-conscious enthusiasts, the Grand Highlander compensates with features like adaptive dampers for smooth off-roading and a versatile trim lineup to satisfy diverse needs. Ultimately, it's the chariot of choice for large families, frequent road trippers, and outdoor adventurers who prioritize space, power, and comfort.

Beyond the Labels: Deciding Factors for Your Perfect Hybrid SUV

Choosing the champion isn't a binary dance of labels. Your ideal hybrid SUV might lie somewhere in the nuanced spectrum between the CR-V's urban prowess and the Grand Highlander's adventurous spirit. Consider these potential deciding factors:

- **Budget:** Can you comfortably afford the higher starting price and potentially higher fuel costs of the Grand Highlander, or does the CR-V's affordability win you over?
- **Driving Habits:** Do you primarily navigate city streets, or do you yearn for weekend escapes

into the wild? Prioritize maneuverability and fuel efficiency for urban adventures, or choose power and space for open-road excursions.

- **Family Needs:** How many passengers do you need to accommodate regularly? Will the CR-V's cabin feel cramped for your family, or does the Grand Highlander's spaciousness sway your decision?

- **Feature Preferences:** Which features are most important to you? User-friendly technology and compact size might allure some, while advanced driver-assistance systems and off-road capabilities might win over others.

Remember, the perfect hybrid SUV is a reflection of your own lifestyle and priorities. Don't hesitate to test drive both vehicles, delve into detailed comparisons, and embrace your gut feeling. This guide has equipped you with the knowledge and insights to make an informed decision, ensuring you choose the hybrid SUV that seamlessly integrates with your life, your adventures, and your environmental conscience.

Claim Your Throne: Finding Your Perfect Hybrid SUV in the Arena of Adventures

The 2024 Honda CR-V Hybrid and Toyota Grand Highlander Hybrid stand poised, not as adversaries, but as champions of distinct driving experiences. While both possess hybrid prowess, their strengths dance to the rhythm of different lifestyles. So, buckle up, adventurers, for this guide crowns the perfect hybrid SUV for your unique journey.

The City Dweller's Triumph: CR-V Hybrid, Urban Eco-Warrior

Scenario: Your domain is the urban jungle, paved streets your battleground. Fuel efficiency is your war cry,maneuverability your weapon. Enter the CR-V Hybrid, the agile warrior of city streets. Its compact size is your cloak,weaving through traffic with nimble ease. Parking becomes a victory dance, not a frustrating struggle. And at

40 mpg combined, your wallet and the planet celebrate as you conquer mile after mile. User-friendly technology keeps you connected, while the comfortable cabin becomes your haven between urban escapades.

Expert Acclaim: "The CR-V Hybrid shines in the city, with its fuel efficiency and nimble handling making it a joy to drive." - Car and Driver

Poll Time: Are you a solo adventurer or a city family of four? Does fuel efficiency top your list, or is compact size your driving force? Choose your CR-V destiny:

- **City Slicker:** Prioritize fuel efficiency and maneuverability?

The base LX trim offers everything you need to rule the urban maze.

- **Tech Savvy Warrior:** Crave the latest gadgetry? Upgrade to the EX-L for a panoramic moonroof and heated seats.

Call to Action: Claim your urban throne! Schedule a test drive and experience the CR-V Hybrid's agility firsthand. Visit your local Honda dealer or explore their website to discover financing options and tailor your urban chariot.

The Grand Highlander's Ascension: Spacious Adventurer's Dream

Scenario: Your spirit craves open roads, your family demands comfort, and adventure whispers tales of distant horizons.The Grand Highlander Hybrid answers the call, a behemoth of space and power ready to conquer any journey. Seven passengers revel in its cavernous interior, while the engine tackles mountain passes with a mighty roar. Advanced technology keeps everyone entertained, even as the miles melt away. Sure, its fuel efficiency might not ignite eco-warrior chants, but the Grand Highlander compensates with off-road capabilities and a versatile trim lineup that caters to diverse needs.

Expert Insight: "The Grand Highlander Hybrid is a game-changer for large families who love road trips. It offers the space, power, and features to make every journey an adventure." - Edmunds

Quiz Time: Do you conquer trails with your tribe, or explore hidden corners with your family? The Grand Highlander awaits your command:

- **Road Trip Royalty:** Need maximum space and creature comforts? The Limited trim with its premium leather seats and panoramic moonroof makes every journey luxurious.
- **Off-Road Adventurer:** Crave rugged escapades? The XLE with its all-wheel drive and enhanced suspension lets you explore beyond the pavement.

Call to Action: Ascend to your adventure! Visit your local Toyota dealer to experience the Grand Highlander Hybrid's spaciousness and power firsthand. Test drive different trims and explore financing options to craft your perfect adventure chariot.

Beyond the Crowns: A Realm of Individual Choice

Remember, these recommendations are your guide, not your master. Don't hesitate to:

- **Test drive both vehicles:** Feel the CR-V's urban agility and the Grand Highlander's spacious comfort. Listen to your gut; it rarely steers you wrong.

- **Dive into comparisons:** Check online reviews and ratings for detailed evaluations of features, performance, and reliability.
- **Consider your finances:** Factor in fuel costs, insurance, and maintenance expenses. Choose an SUV that fits your budget without compromising your needs.

In the end, the perfect hybrid SUV isn't about labels or crowns. It's about finding the champion that seamlessly integrates with your life, your budget, and your wanderlust. So, explore, compare, and test drive. Your dream hybrid SUV awaits,ready to fuel your adventures and crown you the master of your own journey.

Embrace the Open Road:

This guide is just the first step. Keep the conversation going by:

- **Building an online community:** Create a forum for readers to share their hybrid SUV experiences, ask questions,and offer advice to each other.

- **Partnering with dealerships:** Organize test drive events where readers can experience both vehicles firsthand.

- **Sharing updates and news:** Keep your audience informed about new hybrid SUV releases, advancements in technology, and changes in fuel costs or warranty coverage.

Beyond the Horizon: Gazing into the Crystal Ball of Hybrid SUVs

The 2024 Honda CR-V Hybrid and Toyota Grand Highlander Hybrid stand as shining stars in the ever-evolving hybrid SUV landscape. But the automotive realm is a dynamic one, constantly shifting with technological advancements and evolving consumer preferences. So, let's peer into the crystal ball and imagine the future of these two champions,envisioning potential changes and market trends that might reshape the hybrid SUV scene in the years to come.

CR-V Hybrid: Embracing Electric Evolution

- **Electrification Amplification:** Whispers of a fully electric CR-V are growing louder. Honda's commitment to electrification hints that a pure electric CR-V might join its hybrid sibling in the near future, offering urban dwellers an emission-free option while retaining the beloved size and practicality.

- **Connected Cockpit Enhancements:** Expect the CR-V's already user-friendly tech to get even smarter. Advanced driver-assistance features like semi-autonomous driving could find their way into higher trims, while augmented reality displays might enhance navigation and provide real-time traffic updates.

- **Personalized Performance:** Imagine adjusting your CR-V's driving dynamics on the fly. Adaptive suspension systems and customizable drive modes could become mainstream, allowing you to switch between a nimble city cruiser and a sportier weekend warrior with the touch of a button.

Grand Highlander Hybrid: Carving the Path to Adventure

- **Hybrid Powerhouse Unleashed:** While the current Grand Highlander Hybrid offers impressive power, future iterations might push the boundaries with plug-in hybrid technology. Imagine extended electric range for city commutes, seamlessly transitioning to hybrid power for longer journeys, offering the best of both worlds for eco-conscious adventurers.

- **Rugged Refinement Refined:** Expect the Grand Highlander's off-road capabilities to get even more impressive.Advanced all-wheel drive systems, enhanced traction control, and even dedicated off-road driving modes could cater to hardcore adventurers seeking to go beyond paved roads.

- **Family Entertainment Amplified:** Picture your Grand

Highlander becoming a rolling entertainment hub. Rear-seat entertainment systems with enhanced connectivity and immersive audio experiences could keep even the most restless travelers engaged, transforming long journeys into road trip memories.

Market Trends: The Driving Forces of Change

Beyond these specific models, the future of hybrid SUVs will be shaped by broader market trends:

- **Sustainability Surge:** Consumer demand for eco-friendly options is expected to continue its upward trajectory.Expect to see more affordable hybrid and electric SUVs entering the market, catering to budget-conscious drivers.

- **Tech Takeover:** Advanced driver-assistance features will become increasingly sophisticated, paving the way for semi-autonomous and even fully autonomous driving in the future. Look for hybrid SUVs seamlessly integrating these technologies, revolutionizing the way we navigate our roads.

- **Personalization Reigns:** Customization will become key. From adjusting ride height and suspension settings to tailoring driving modes and even choosing interior accents, expect hybrid SUVs to offer more ways to reflect your individual preferences.

A Future Awaits: Choosing Your Path in the Hybrid Arena

While predicting the future is never an exact science, these glimpses into the world of hybrid SUVs offer a glimpse of the exciting paths that the CR-V and Grand Highlander might take. Remember, choosing your ideal hybrid SUV isn't just about the present; it's about choosing a companion that can evolve with your needs and adapt to the changing landscape of the automotive world.

Consider these questions as you gaze into the hybrid horizon:

- **Do you prioritize fuel efficiency and urban agility, or crave spaciousness and adventurous capabilities?**
- **How important are cutting-edge tech features and advanced driver-assistance systems to you?**
- **Are you willing to embrace potential future updates and electric iterations of your chosen hybrid SUV?**

By answering these questions and staying informed about market trends, you can choose a hybrid SUV that not only complements your current lifestyle but also possesses the potential to grow and adapt alongside you on the ever-evolving road of automotive advancements.

Appendix: Demystifying the Details

Your journey towards the perfect hybrid SUV doesn't end with the final verdict. This appendix equips you with the intricate details and technical know-how to truly understand the inner workings of the 2024 Honda CR-V Hybrid and Toyota Grand Highlander Hybrid. Dive deep into the specifications and unravel the mysteries of technical terms, ensuring you make an informed and confident choice.

Honda CR-V Hybrid: A Compact Powerhouse

Specifications:

- **Engine:** 2.0L 4-cylinder Atkinson-cycle engine + electric motor
- **Combined System Horsepower:** 215 hp
- **Transmission:** Electronically controlled continuously variable transmission (eCVT)
- **Fuel Economy:** 40 mpg combined (EPA estimate)
- **Seating Capacity:** 5 passengers
- **Cargo Space:** 39.2 cubic feet behind rear seats, 75.8 cubic feet with seats folded
- **Safety Features:** Standard Honda Sensing® suite including Collision Mitigation Braking System™, Lane Departure

Warning, and Adaptive Cruise Control with Low-Speed Follow.

- **Warranty:** 3-year/36,000-mile bumper-to-bumper warranty, 5-year/60,000-mile powertrain warranty, 8-year/100,000-mile hybrid battery warranty.

Glossary of Terms:

- **Atkinson-cycle engine:** A type of engine designed for increased fuel efficiency by optimizing the intake and exhaust cycles.
- **eCVT:** A transmission that uses a combination of gears and electric motors to smoothly transfer power from the engine to the wheels.

- **EPA estimate:** The official fuel economy rating determined by the Environmental Protection Agency.

Toyota Grand Highlander Hybrid:
Spacious Adventure Seeker

Specifications:

- **Engine:** 2.5L 4-cylinder Atkinson-cycle engine + electric motor
- **Combined System Horsepower:** 296 hp
- **Transmission:** 8-speed automatic transmission
- **Fuel Economy:** 29 mpg combined (EPA estimate)
- **Seating Capacity:** 7 passengers (8 passengers optional)

- **Cargo Space:** 16.0 cubic feet behind third-row seats, 48.4 cubic feet behind second-row seats, 84.3 cubic feet with all seats folded
- **Safety Features:** Standard Toyota Safety Sense® 2.0 suite including Pre-Collision System with Pedestrian Detection, Lane Departure Alert with Steering Assist, and Automatic High Beams.
- **Warranty:** 3-year/36,000-mile bumper-to-bumper warranty, 5-year/60,000-mile powertrain warranty, 8-year/100,000-mile hybrid battery warranty.

Glossary of Terms:

- **Atkinson-cycle engine:** Same as above.

- **8-speed automatic transmission:** A traditional automatic transmission with eight gears for smoother acceleration and improved fuel efficiency.

Beyond the Numbers: Understanding the Nuances

Remember, specifications are just one piece of the puzzle. Consider these additional factors when evaluating your ideal hybrid SUV:

- **Driving Feel:** Test drive both vehicles to experience their handling, acceleration, and overall driving experience.

- **Technology and Features:** Compare the infotainment systems, driver-assistance features, and available comfort amenities to see which best suit your needs.

- **Budget and Ownership Costs:** Factor in fuel costs, insurance, and potential maintenance expenses to determine the long-term affordability of each vehicle.

Personalize Your Appendix:

To enrich your guide, consider these additions:

- **Comparison tables:** Create side-by-side comparisons of key specifications for easy reference.

- **Interactive features:** Include quizzes or calculators to help readers determine their fuel costs or compare warranty coverage.
- **Expert insights:** Quote car reviewers or analysts on the strengths and weaknesses of each vehicle's technical aspects.

By combining detailed specifications, a user-friendly glossary, and insightful analysis, you can transform this appendix from a dry collection of numbers into a valuable resource that empowers your readers to make informed and confident decisions about choosing their perfect hybrid SUV.